RAIN FOREST

A Science Discovery Book

by Annalisa McMorrow
illustrated by Marilynn G. Barr

Dedicated to Lisa Levine

Publisher: Roberta Suid
Design & Production: Scott McMorrow
Cover Design: David Hale
Cover Art: Mike Artell

Other books of interest:
Save the Animals! (MM 1964), *Love the Earth!* (MM 1965), and *Learn to Recycle!* (MM 1966)

Entire contents copyright ©1999
by Monday Morning Books, Inc.

For a complete catalog, please write to the address below:
P.O. Box 1680
Palo Alto, CA 94302

Call our toll-free number: 1-800-255-6049
E-mail us at: MMBooks@aol.com
Visit our Web Site:
http://www.mondaymorningbooks.com

Monday Morning Books is a registered trademark of Monday Morning Books, Inc.

Permission is hereby granted to reproduce student materials in this book for non-commercial individual or classroom use.
ISBN 1-57612-079-1
Printed in the United States of America
98765432

CONTENTS

A Note to Teachers and Parents	4
Welcome to the Rain Forest	5
Rain Forest Quiz	7
Answers to the Quiz	8
Scientist's Notebook	9
More About Rain Forests	10
Rain Forest World Map	12
Rain, Rain, Here to Stay!	14
Rain and Rainbows	15
Monkeying Around	16
How Loud Are They?	17
Sticky Frogs	19
Flying Snakes	21
Add Them Up	22
As Slow As a Sloth	23
Rain Forest Fables	24
Goldilocks and the Three Gorillas	25
Mother Goose of the Jungle	26
Spot the Camouflage	27
Giant Creepy-Crawlies	29
You're a Grand Old Toucan	31
Stay Away!	33
Air Plants	35
Miraculous Mushrooms	36
Create a Canopy	37
Amazing Ants	38
Name the Animals	41
Save the Rain Forests!	43
Rain Forest Trading Cards	45
Rain Forest Glossary	47
Rain Forest Resources	48

A NOTE TO TEACHERS AND PARENTS

This book is filled with exciting activities for children. Each page is written directly for the children, with activities and information that they can easily understand.

The activities in *Rain Forest* cross the curriculum: math, language arts, science, and art. For example, children compare a variety of household objects with the length of different rain forest creatures (like tiger centipedes); they locate tropical rain forests on a map; they make a meal fit for a gorilla; they write their own rain forest tongue twisters; they learn about similes, and much more!

If you are working through this book as a class, simply photocopy the activities and give a page to each child (or each group of children, if you are doing group activities).

Note that many of the projects are perfect for homework assignments. If you are teaching about geography, you can have children study the Rain Forest World Map and then do research to find out different facts about the locations of rain forests. Children can then bring back their information to share with the class.

Parents can also help their children with the activities in this book. You can read the directions aloud and do the activities together. For example, parents and children can chart the rainfall in their community together. You can compare the rainfall your area receives with the amount that falls in a tropical rain forest. You can also learn about past records and see how close this year's rainfall come to breaking them!

The most important thing about *Rain Forest* is that children will be having fun while they explore the scientific world!

WELCOME TO THE RAIN FOREST

Tropical rain forests are fascinating places! Many amazing animals live in rain forests. Although these animals range from tiny mites to ferocious jaguars, the creatures have one major thing in common. They get wet a lot! Rain falls nearly every day in tropical rain forests. In fact, there is usually at least 60 inches (152 cm) of rain per year. Often there's even more.

Before you start learning about tropical rain forests, you're going to need a passport. That's because tropical rain forests are found around the world on different continents.

Either draw a picture of yourself or tape a photograph of yourself in the passport pattern. Then cover your picture (or drawing) with clear contact paper to protect it.

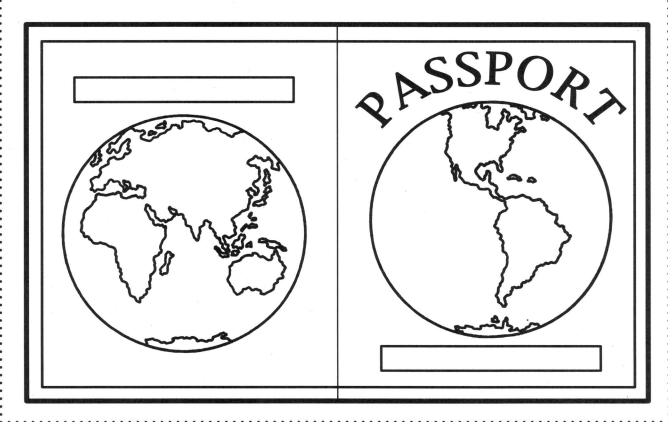

Now that you have a passport, you're ready to go! But before you learn all about tropical rain forests, check your current rain forest knowledge with the Rain Forest Quiz. Then check the Answers to the Quiz. Finally, look at the "Rain Forest World Map." This will show you where different kinds of rain forests are located.

Throughout the activities in this book, you can use your "Scientist's Notebook" to start recording information you've learned. Scientists often keep notebooks to help them remember what they've learned. They write down facts, jot notes to themselves, or even draw pictures of what they've seen. Whenever you want to remember a specific bit of information, write it in your notebook. If you run out of room, use any lined paper.

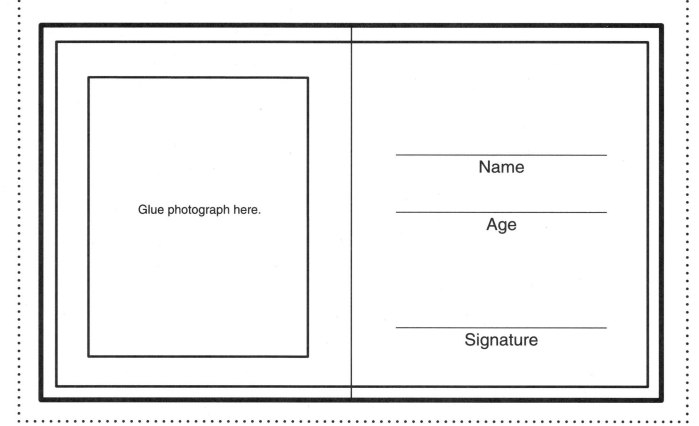

RAIN FOREST QUIZ

1. The tall trees in the rain forest have very deep roots.

2. Few animals live in a rain forest because it's so wet.

3. Rain forest trees can be 200 feet (60 meters) tall.

4. Army ants living in the rain forest travel in swarms of up to 20 million ants.

5. The black and orange tiger centipede is more than 9 inches (23 cm) long.

6. Some flowers in the rain forest smell like rotting meat.

7. In most rain forests it rains more than 200 days a year.

8. If all the rain forests in the world were put together, they would be about the size of ten football fields.

9. A type of cockroach in the rain forest is so large it would fill the palm of your hand.

10. The gorilla's diet consists almost entirely of plants.

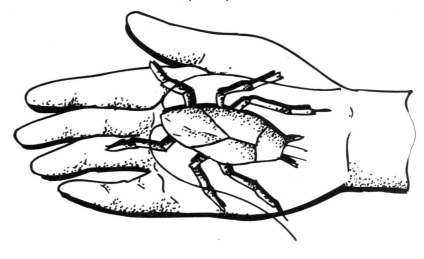

ANSWERS TO THE QUIZ

1. The tall trees in the rain forest have very deep roots. FALSE! Trees in the rain forest have shallow roots. That's because the soil isn't rich. The roots take in the nutrients near the surface.

2. Few animals live in a rain forest because it's so wet. FALSE! Tropical rain forests are filled with different mammals, birds, reptiles, and amphibians.

3. Rain forest trees can be 200 feet (60 meters) tall. TRUE! These trees tower over the top of the rain forest, which is called the canopy.

4. Army ants living in the rain forest travel in swarms of up to 20 million ants. TRUE! These ants march along the forest floor in search of food.

5. The black and orange tiger centipede is more than 9 inches (23 cm) long. TRUE! The centipede's bright black and orange colors tell other animals that it is poisonous.

6. Some flowers in the rain forest smell like rotting meat. TRUE! Orchids often have strong scents. Some smell like perfume, others like rotting meat.

7. In most rain forests it rains more than 200 days a year. TRUE! In one year, 240 inches (609 cm) of rain can fall.

8. If all the rain forests in the world were put together, they would be about the size of ten football fields. FALSE! If all the rain forests were put together, they would be about the size of the United States!

9. A type of cockroach in the rain forest is so large it would fill the palm of your hand. TRUE! This cockroach is called the Blaberus giganteus.

10. The gorilla's diet consists almost entirely of plants. TRUE! Gorillas mainly eat leaves.

SCIENTIST'S NOTEBOOK

More About Rain Forests

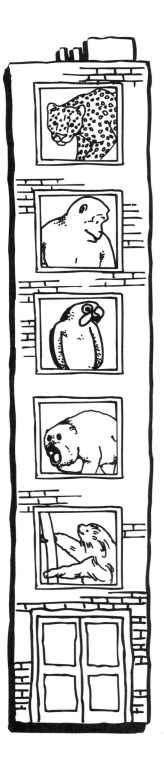

Picture a tall building. Different things happen on different stories of the building. A rain forest is sort of like that. It is made up of several layers. Different types of plants and animals live in each layer.

The top of the rain forest is called the emergent layer. This layer consists of the tops of the tallest trees in the rain forest. These trees are scattered apart from each other. They can be 200 feet (60 m) tall!

The canopy layer of the rain forest is made up of trees that are usually 60 to 90 feet (18 to 27 m) tall. These trees grow close together. The top of the canopy in most rain forests gets a lot of sun. The branches of the canopy block the sun from reaching the forest's lower levels.

The understory is the area from near the ground up to 40 or 50 feet (12 to 15 m) in the air. This part of the rain forest contains the trunks of the canopy trees, young trees still growing toward the canopy, and trees that can live their whole lives in the shade. Shade-tolerant trees get very little light, but they can survive without it.

The lowest level of the rain forest is called the forest floor. The soil in the rain forest is not very rich. Most of the nutrients are found on top of the ground. Because of this, the trees in the rain forest have shallow roots. Their roots don't go deep into the earth. This makes the trees less sturdy. (Shallow roots aren't as supportive as deep roots.) Some trees develop supports that spread out from the base. Called buttresses, these supports help keep the trees from falling over!

PICTURE A RAIN FOREST

RAIN FOREST WORLD MAP

Tropical rain forests can only exist in warm, wet areas. Most are found near the equator.

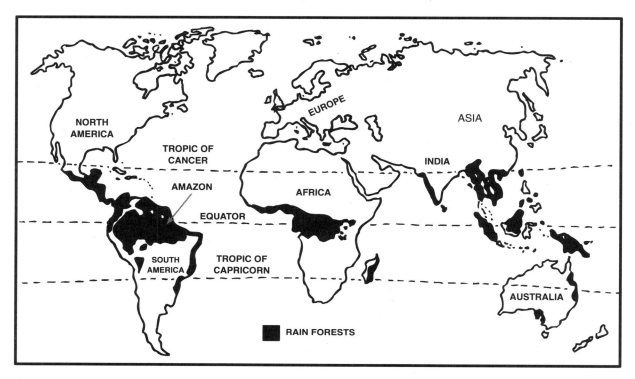

As you can see, tropical rain forests exist in Central and South America, India, Southeast Asia, the Philippines, northeastern Australia, and the Congo River basin in Africa. Find the rain forest that's nearest to you!

Just for Fun: Visit a Forest

In the past, when people traveled they collected stickers to put on their suitcases to show the places they'd been. Imagine that you traveled to the different rain forests. Wouldn't you want labels for your suitcase? Trace or cut out the Rain Forest Luggage Labels to decorate your binder or book covers. Then use the blank to make one of your own.

RAIN FOREST LUGGAGE LABELS

I Left My Leaf in a Central American Rain Forest!

I'M STUCK ON THE RAIN FORESTS IN AUSTRALIA.

I hung around in a South American Rain Forest.

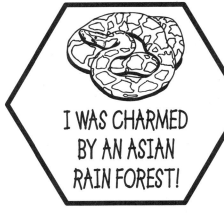

I WAS CHARMED BY AN ASIAN RAIN FOREST!

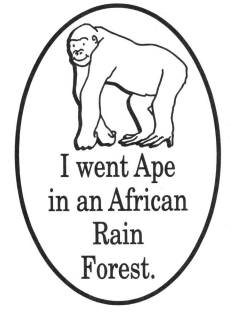

I went Ape in an African Rain Forest.

Rain Forest © 1999 Monday Morning Books, Inc.   13

RAIN, RAIN, HERE TO STAY!

Tropical rain forests can receive from 70 to 240 inches (178-610 cm) of rain every year. In most tropical rain forests, it rains more than 200 days a year. How much does it rain in your area? Read the weather reports in your local newspaper, or look at weather reports online to find out how much rain your area usually receives. Then follow the weather reports during your rainy season and see how much rain your area receives. Keep checking! Do you get as much rain as a tropical rain forest? How close do you come?

The rainfall report in your local paper will look something like this:

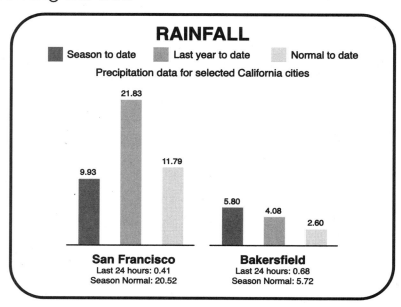

Just for Fun: How Hot Is It?

In tropical rain forests, the temperature rarely falls below 70 degrees Fahrenheit (21 degrees Celsius). When the sun shines, temperatures can reach 90 degrees Fahrenheit (32 degrees Celsius). Compare the temperatures in your area with these temperatures. How close does your climate come to that of a tropical rain forest?

RAIN AND RAINBOWS

Not everyone likes rain, but most people like rainbows. Rainbows happen when light shines through water droplets in the air. Sunlight contains all the colors in a rainbow. When sunlight enters a raindrop, the light rays bend. The colors split apart so that you can see each one.

Can you have a rainbow without rain? Yes, you can! Next time you see sprinklers on a sunny day, try to find the rainbow in the water. Or make a rainbow yourself. All you'll need is a mirror, a pan of water, and a light source.

Put the pan of water in a sunlit place opposite a white wall. Hold the mirror at one end of the pan so that the sunlight strikes it. Move the mirror until you see rainbow colors on the wall.

Draw a picture of what you see in your Scientist's Notebook. Keep track of the different rainbows you see after it rains. Look for a double rainbow. It appears outside the main rainbow. Its colors are paler. Instead of following the normal order of red, orange, yellow, green, blue, indigo, and violet, the colors appear in reverse.

Just for Fun: Rainbows in Flight

A wide variety of colorful birds live in the rain forest. When they fly through the air, it's like a rainbow in flight! Make colorful parrots from colored construction paper. (You can trace the bird pattern in the margin, or cut out birds of your own.) Glue on colored tissue paper for feathers, or use colored feathers (often available at crafts stores). Glue the birds in rainbow color order onto a construction paper background.

Monkeying Around

All of the world's apes, and most of its monkeys, live in the tropical rain forests. Different types of apes and monkeys live in different tropical rain forests. Orangutans live on the southeast Asian islands of Sumatra and Borneo. Chimpanzees are found in central and west Africa. Gorillas live only in deep rain forests in central Africa.

The male gorilla has a silver back, can reach 6.5 feet tall (1.8 m), and weigh more than 600 pounds (272 kg)! Although male gorillas look fairly ferocious, they only fight if they have to protect their families. How many of your friends would it take to equal the weight of a male gorilla?

First, weigh yourself and write down the weight in your Scientist's Notebook. Then weigh a friend and write your friend's weight below yours. Add the two weights together. Continue weighing your friends and adding their weights until you reach 600 pounds (272 kg). It's okay if you're a little bit under or a little bit over.

Fun Fact: Apes can swing hand over hand. Monkeys can't. Monkeys run on branches on all fours.

Just for Fun: Eat Like a Gorilla

Gorillas may look scary, but they are actually shy and gentle. They eat nuts, berries, and fruits. Put together a fruit salad that has some of each of these ingredients. You might add a banana, since many monkeys enjoy them! Sprinkle nuts on top of your fruit salad and share it with your family.

How Loud Are They?

Some monkeys are very noisy. They make noise to warn other monkeys to stay away from their food supply. One type of monkey, called the howler monkey, can be heard up to 2 miles (3 km) away. Gibbons make shrieks and barks that can be heard up to .6 mile (1 km) away.

How far is that? Next time you take a ride in a car, have the driver set the odometer to zero. Ask the driver to tell you when you are .6 mile (1 km) away from your house. This is how far a gibbon's voice travels. Have the driver tell you next when you are 2 miles (3 km) away. If a howler monkey was howling from your house, you'd still be able to hear it!

Just for Fun: More About Monkeys

Use the ape and monkey patterns to test your friends' knowledge about these interesting creatures. First, cut out the patterns. On one side of each pattern, write a question about apes or monkeys. On the other side, write the answer. You can use facts you already know, or do some research on your own.

When you're finished, hang a piece of green string across your room and fasten the monkeys to the string with clothespins. Then test your friends and family to see how much they know about monkey business!

APES AND MONKEYS

STICKY FROGS

Can you imagine living your whole life without setting foot on the ground? Some tree frogs do! They spend their entire lives in the canopy of the rain forest, never going down to the ground! How do the frogs climb around on slippery leaves and branches? They have sticky pads on their fingers and toes!

Imagine that you had sticky pads on your hands and feet. (They could be special gloves and shoes you could put on, or they could be an actual part of you!) Write a short story about what it might be like to live for one day with these pads. What would you do to surprise your friends? How would you be a better athlete? What would be good and bad about having sticky feet and hands?

When you're finished with your story, illustrate it and share it with your friends and family.

Just for Fun: Funky Frogs

Use the Funky Frogs to make your own sticky frogs. Cut out the frogs and color them however you desire. Then use two-sided tape to stick them wherever you want. You might use the sticky frogs to decorate your binder, your bedroom, or wherever you'd like to put a fun frog friend.

FUNKY FROGS

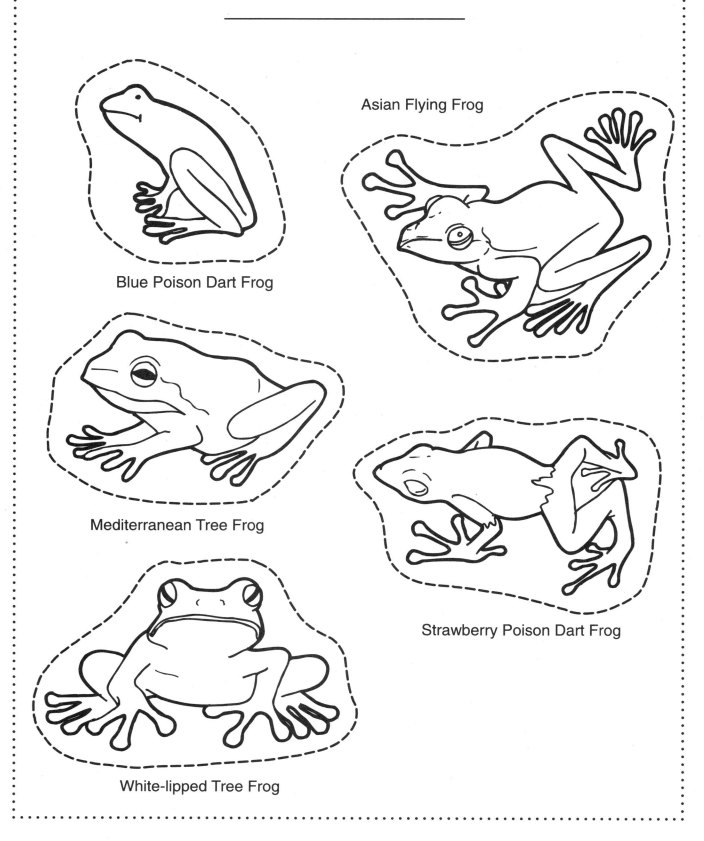

Blue Poison Dart Frog

Asian Flying Frog

Mediterranean Tree Frog

Strawberry Poison Dart Frog

White-lipped Tree Frog

FLYING SNAKES

Look, up in the sky. It's a bird! It's a plane! It's ... a flying tree snake! In Southeast Asia, there are several types of snakes that seem to fly. In fact, they glide through the air from tree to tree. How do they do it? These snakes raise their ribs upward and outward as they travel. This helps them to flatten their bodies in the air. A flying snake can travel up to 165 ft (50 m) from one tree to another.

With your friends, measure this distance. Use measuring sticks or tape measures, keeping track as you go along. Compare the distance to other objects that you're familiar with. For example, how many desks would it take to fit this distance? (Measure your desk, then divide 165 by the number of feet your desk is long.) Or how many classrooms would it take to equal 165 feet? Measure the length of your classroom. Then divide 165 by the number of feet your classroom is long.

Just for Fun: Snake Arm Band

Some of the snakes found in the rain forest are very poisonous. You'd have to avoid them on a visit to the rain forest. However, not all people consider snakes unlucky creatures. In Egypt, snakes were sometimes featured on jewelry. People even wore bracelets and arm bands in the shape of snakes. Make your own arm band by forming a piece of aluminum foil into a tube. Add eyes and a mouth using markers. Then coil the foil snake around your arm! Or draw a picture of a snake for your arm band.

ADD THEM UP

The rain forest is a busy place. A typical patch of rain forest covering 4 square miles (10.4 square kilometers) contains 750 species of trees, 750 species of other plants, 125 species of mammals, 400 species of birds, 100 species of reptiles, and 60 species of amphibians! That's a lot of different creatures!

Next time you walk to school, or walk in a park, or walk in your neighborhood, keep track of how many different plants and animals you see. Make a column for plants, trees, mammals, birds, reptiles, and amphibians. If you don't know the names of the plants or animals you see, draw pictures in your Scientist's Notebook. See if you can find them later in a book about local wildlife.

Do you come close to the amount of wildlife found in the rain forest?

Just for Fun: Learn About Your Area

Check out a book about the birds, flowers, or trees that live in your area. Try to find as many of the listed birds or plants as you can. Keep notes in your Scientist's Notebook.

AS SLOW AS A SLOTH

Sloths are creatures that live in the rain forest. Two kinds of sloths live in Central and South American rain forests. One type has three toes. The other type has two toes.

A sloth's fur often looks green. This is because algae grows on it. The green tint helps the sloth to blend in with the leaves.

Sloths are slow creatures. In general, they don't move very much. They are so slow that it can take them more than a day to move from one tree to another! The word "sloth" is sometimes used to describe someone who is slow-moving or sluggish. A person who is extra tired might say, "Today, I'm as slow as a sloth."

This is called a simile. Similes are figures of speech in which one thing is compared to another, dissimilar thing by the use of the words *like* or *as*. Make a list of as many similes as you can think of based on creatures or plants that live in a rain forest. For instance, since flying snakes can travel a great distance in one leap, you might say that a basketball player flew across the court like a flying snake! Keep track of all your rain forest similes in your Scientist's Notebook. Try to think of similes for a turtle, a toucan, and a jaguar!

Just for Fun: Sleepy Sloths Slink Slowly

Tongue twisters often use words that start with the same letter or letters. Tongue twisters may alternate words that start the same way, for example: She sells seashells by the seashore. (The s- and sh- sounds alternate.)

Try to create tongue twisters using different creatures or plants that live in the rain forest, such as an orangutan, a jaguar, a centipede, or a parrot. When you've written your tongue twisters, illustrate them and bind them in a book. Trade tongue twisters with your friends.

RAIN FOREST FABLES

Aesop's fables are tales that use animals to teach a lesson. When you finish reading the story, there is a moral at the end. For example, in the fable about the Tortoise and the Hare, the Hare is so sure that he will win the race, he actually takes a nap. The tortoise is slow, but steady. He doesn't stop to rest, he simply moves at his own pace, and he wins. The moral "slow and steady wins the race" means that just because you're fast doesn't mean you'll always come out ahead.

Other Aesop's fables use different animals and have different morals. Choose a fable to rewrite using animals that live in the rain forest. If you were going to rewrite the Tortoise and the Hare, you might choose another slow animal, like the sloth, in place of the Tortoise. For the Hare, you could choose a quick animal, such as a gibbon. (Gibbons swing through the trees from branch to branch, moving quickly through the forest canopy.)

When you're finished writing your fable, illustrate it. Here are some fables you might want to rewrite:
The Fox and the Grapes
The Lion and the Mouse
The Travellers and the Bear

You can find more fables in any book of Aesop's fables. Check your local library for resources!

Just for Fun: Friendly Fables
Work with your friends to create a whole book of retold rain forest fables. Make photocopies so that each person gets to have a copy!

Goldilocks and the Three Gorillas

You probably know several fairy tales by heart, such as Cinderella, Sleeping Beauty, Snow White, Hansel and Gretel, Goldilocks and the Three Bears, Jack and the Beanstalk, and the Goose That Laid the Golden Egg. Think about your favorite fairy tale. Now, imagine that the tale took place in a rain forest. Rewrite the story in a rain forest setting. For example, if you chose to rewrite Goldilocks, you might turn the bears into gorillas. Goldilocks might visit their tree house instead of a regular house.

When you write your version, consider the following questions: How might the ending be different? How might the characters behave if they were surrounded by vines and tall trees and rain that rarely stopped?

Write and illustrate your own fairy tale. Consider working with a friend and writing several. Then bind the stories together in a Rain Forest Fairy Tale book!

Just for Fun: Charlie in the Rain Forest

If you don't want to write a fairy tale, you could also rewrite a famous story. For example, try setting Alice in the rain forest instead of Wonderland or Dorothy in the jungle instead of Oz! Or even Charlie in the rain forest, instead of the chocolate factory. You don't have to rewrite an entire book. Try rewriting your favorite scene, or the opening.

Mother Goose of the Jungle

Although Mother Goose includes rhymes about many different animals—cats, mice, geese, horses, cows, and pigs—there aren't many about rain forest animals. For instance: *Old Mother Hubbard went to the cupboard to fetch her poor dog a bone.* She didn't go to the cupboard to find a snack for her ocelot or her spider monkey! Why should rain forest animals be left out? You can fix this situation by creating your own book of "Mother Goose of the Jungle" rhymes. One simple way to write rhymes is to change words in a poem or rhyme you already know. For example:

Hey, diddle, diddle,	*Hey, diddle, diddle,*
The cat and the fiddle,	*The sloth and the fiddle,*
The cow jumped over the moon.	*The frog jumped over the moon.*
The little dog laughed	*The jaguar laughed,*
To see such a sight,	*To see such a sight,*
And the dish ran away	*And the dish ran away*
With the spoon.	*With the spoon.*

You can write poems to Mother Goose rhymes or any nursery rhymes. If you add facts to your poems, you might have an easier time memorizing the facts. Remember, you don't have to focus only on rain forest animals. You can also write rhymes about rain forest plants! For example:

Mary, Mary, quite contrary,	*Mary, Mary, quite contrary,*
How does your garden grow?	*How does your jungle grow?*
With silver bells and cockle shells,	*With orchids, ferns, and buttresses,*
And pretty maids all in a row.	*And mushrooms all lined in a row!*

Just for Fun: Rain Forest Rhyme Book

After you write several new poems, collect them in a book. Draw a picture for each poem. Make a cover for your book and staple the cover to your pages. You might even have your friends help you. Each one could add a page to the book.

SPOT THE CAMOUFLAGE

Some rain forest creatures are hard to find. This is because they are camouflaged to blend in with their surroundings. Why do animals use camouflage? To hide from their enemies! When these animals stay perfectly still, they're almost impossible to see.

Camouflaging animals include insects that look like sticks, butterflies with wings that look like dried leaves, and frogs that blend in with green leaves. Some of the animals and insects that use camouflage to survive are the sloth, the orchid mantis, and ocelot. (These are pictured at right.)

Imagine that the rain forest creatures lived in your town. Draw a picture of your house or your neighborhood. Then cut out the rain forest creatures and place them on the picture. Blend them in with the surroundings. For instance, a green frog might blend in on a green car. A red insect might blend well on a stop sign. Have your friends try to spot the hidden creatures when you're finished.

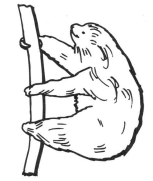

Just for Fun: Camouflaged Bank

The matamata is an Amazonian turtle that looks like a rock. It stays still in shallow water until its dinner comes along. Create a matamata bank to camouflage your money. Use an empty plastic margarine container (or other similarly shaped container). Decorate the outside with papier-mâché and paint. When it's dry, have an adult help you cut a slit in the rounded top. Then, when you want your money, you can take off the plastic lid (which is now your turtle bank's belly).

RAIN FOREST CREATURES

GIANT CREEPY-CRAWLIES

The atlas moth is one of the largest moths. It has a wingspan of 10-12 inches (25-30 cm). The tiger centipede, an orange and black striped centipede, can grow to be over 9 inches (22 cm) long. A curlyhaired tarantula can grow to be 3.5 inches (9 cm) across.

Measure a variety of things around your house or classroom to find items that are the same length as a giant moth, a tiger centipede, and a curlyhaired tarantula. Find several items for each length. Write down the name of each item and its length. Keep track of these items in your Scientist's Notebook.

Giant Moth
10-12 inches (25-30 cm)

Matching Items
Dictionary—11 inches (28 cm)

Just for Fun: Atlas Attraction

The wings of the atlas moths have interesting patterns and colors. This helps the males and females to find each other. Using the moth pattern, design your own set of wings. Choose colors and designs that you find attractive. When you're finished, cut it out and post it on a wall or on a window where the light can shine through.

MAKE-A-MOTH

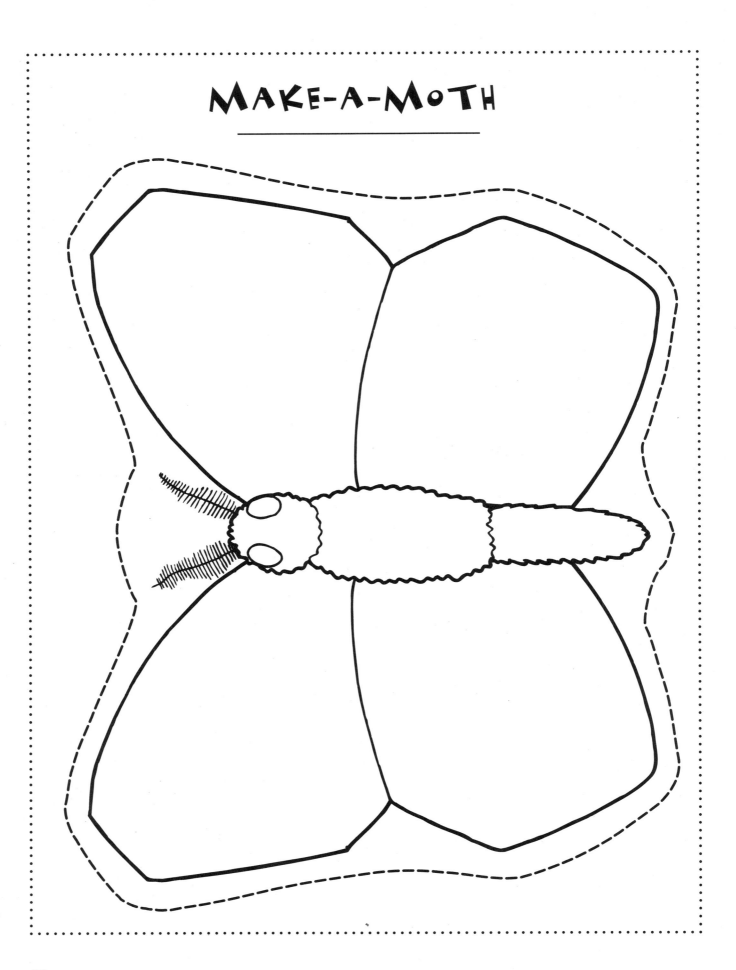

YOU'RE A GRAND OLD TOUCAN

Toucans have brightly colored bills. No two toucans have identical bills. Each bird's bill is slightly different in color and pattern. A Cuvier's toucan has a bill that is yellow, red, orange, and blue. The colors help it locate birds of its own kind and find a mate.

Toucans use their bills like flags to signal other toucans. Using the toucan pattern, make a toucan with a flag-colored bill. The bill can be based on flags from other countries, or from your own. On the back of each bird, write the name of the place where the flag is from. Test your friends to see if they know the flags! If you duplicate the pattern, you can make toucans for different states of the United States of America or Australia, or each province of Canada, and so on.

Just for Fun: Fine-Feathered Report

You can also use the toucans to create a fact-filled report. Write a different rain forest fact on each toucan's bill. Punch a hole in the top of each toucan and thread a piece of yarn through them all. Suspend the toucans overhead where people can see them and read the facts.

FLAG PATTERNS

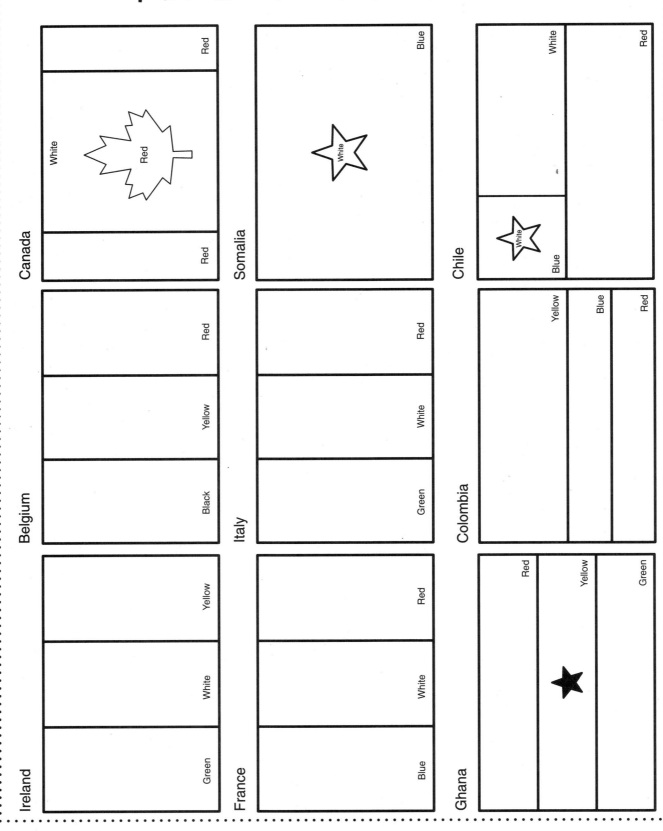

STAY AWAY!

Many colorful animals live in the rain forest. Their colors can act as a warning to other animals. The toucan has a bill that is colored in red, blue, and green. These colors may frighten off predators. Rain forest frogs come in a range of colors. Some of these frogs secrete poisons through their skins. Other animals know to stay away from the frogs based on their colors.

One of the most amazing insects in the rain forest is the passion flower butterfly. Although it looks pretty, this butterfly has a secret. It tastes awful! The passion flower butterfly lays its eggs on leaves of a poisonous passion flower plant. When the eggs hatch, the caterpillars emerge and eat the leaves. These caterpillars are immune to the poison. They store the poison and pass it over to the adult butterflies. With the poison in their systems, the butterflies are protected from birds, because the poison makes them taste bad! (Birds know to stay away because of the coloring of the butterfly.)

Think of things in nature that you know to stay away from based on their colors. (Poison oak. Skunks. Black widows.) Keep a running list in your Scientist's Notebook.

Just for Fun: A Bevy of Butterflies

Create a butterfly mobile using the butterfly patterns. Color them as desired. Poke a hole in each one and tie them to a hanger. Cover the body of the hanger with felt or fabric and hang in your room. If you want to make your mobile more interesting, write a rain forest fact on the back of each pattern. When people admire your mobile, they'll learn something about the rain forest!

BUTTERFLY PATTERNS

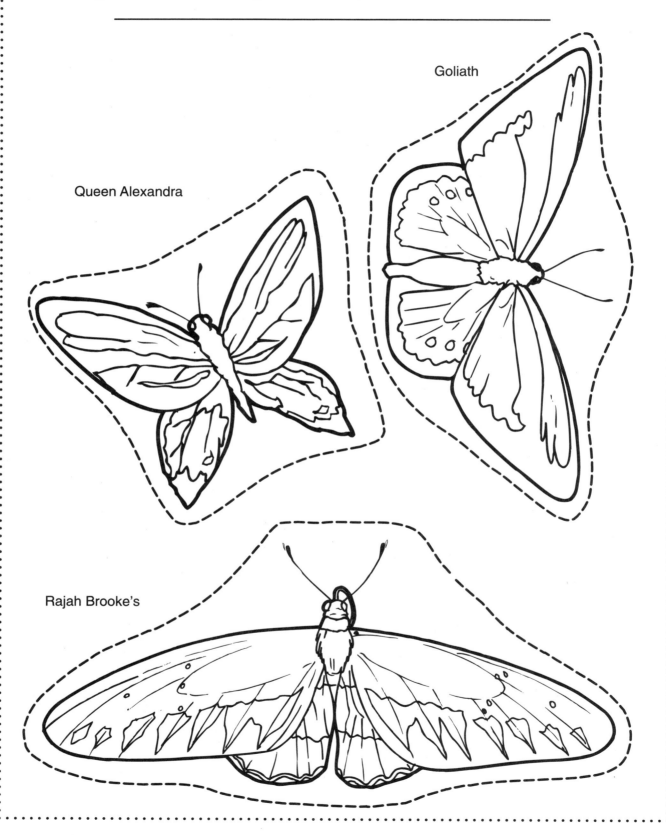

AIR PLANTS

Epiphytes are plants that live in the canopy of the rain forest. Unlike most plants, they don't have roots that go into the ground. Instead, these plants perch on trees and other plants. Epiphyte plants live off nutrients in the air. They take in the nutrients through their tissues or through roots that hang down. Epiphytes are also called air plants.

Air plants often have beautiful flowers. Their flowers attract insects, birds, and other animals that feed on their nectar. Some types of orchids are epiphytes. A single orchid plant may produce up to a million seeds!

There are many different types of air plants, including the ones shown in the margin. The plant on top is a type of orchid. The one shown below is an epiphytic fern. If you go to a botanical garden or a garden supply center, see if you can find any epiphytes!

Just for Fun: An Array of Air Plants

Using colorful tissue paper and green pipe cleaners, make several flowers. You can simply gather the tissue paper into a flower shape and bind the bottom with part of the pipe cleaner. Use the free end of the pipe cleaner stems to hang the flowers. Hang them from door handles, drawer knobs, window sills, and so on.

MIRACULOUS MUSHROOMS

While epiphytes are an unusual plant near the top of the rain forest, mushrooms are an unusual plant near the bottom! Very little light reaches the forest floor. That's because the canopy is so complete, it blocks out most of the light. The plants that grow on the rain forest floor must be able to grow without light. These plants are mostly fungi, or mushrooms. Some of these plants actually glow in the dark!

Mushrooms are interesting plants. They tend to grow where it is wet and damp. They can come in many different colors, including red, yellow, purple, green, orange, and brown.

People used to believe that mushrooms grew in rings to show where fairies had danced. These were called fairy rings. Make your own fairy ring of mushrooms by cutting out mushroom shapes from construction paper. Put them in a ring on a larger piece of construction paper. Write one fact about mushrooms—or about the rain forest—on each mushroom. You can use the facts found on this page, or do your own funky fungi research!

Important Note!
Remember, NEVER eat a mushroom you find growing. Some mushrooms are poisonous. Only eat mushrooms that you buy at the grocery store!

CREATE A CANOPY

Have you ever seen a bunch of people walking with umbrellas on a rainy day? When the umbrellas get close together, the rain can't hit the ground, it just rolls over the top. The canopy in the rain forest works almost the same way. The trees in the rain forest grow so close together that their branches and leaves form a sort of umbrella. The rain can only reach the ground by rolling down the tree trunks.

The rain forest canopy provides shade for the rain forest floor. It also serves as home to many plants, insects, birds, and animals. Some of the animals, like the tree frog, never leave the canopy at all. They spend their whole lives in the canopy trees!

Create a rain forest canopy using a cardboard box. Paint the inside of the box using green and brown tempera paint for a backdrop. Use tubes (like toilet paper tubes or paper towel rolls) to form tree trunks. Make the leafy tops from green construction paper, and use brown pipe cleaners to form the branches. Anchor the trees with tape or glue.

Once you've made a canopy, add the birds, insects, flowers, and animals that live there. Make snakes by coiling green pipe cleaners around the branches. Use brown and black felt to make fruit bats. Add colorful tissue paper flowers. Make tree frogs, monkeys, woodpeckers, and gibbons from colorful clay.

Just for Fun: Rain Forest Diorama

Complete the rain forest diorama by adding the different types of animals and plants that exist at each level of the rain forest: the emergent level, the canopy, the understory, and the forest floor. Birds fly in the emergent level. Opossums, silky anteaters, and others live in the canopy. Kinkajous, spider monkeys, leaf butterflies, and other creatures live in the understory, along with vines that creep up the tree trunks. Moss, fungi, herbs, and insects live on the forest floor.

AMAZING ANTS

Many amazing ants live in the rain forest. Army ants, azteca ants, and leaf-cutter ants are just a few of the types of ants that consider the rain forest home. Leaf-cutter ants are unusual. They snip pieces of leaves from different plants and bring the leaves back to their nests.

Army ants are also interesting. They march in columns that may be 40-feet (12-m) wide in front, eating anything they come upon. This includes snakes, lizards, and insects. The swarm moves at about 65 feet (20 m) an hour, devouring animals they find!

Azteca ants are protectors of trees. They live in the cecropia trees of Latin American rain forests. Whenever their tree is disturbed, the ants emerge from the tree to bite and sting the enemy.

You can see a wide range of ants in the rain forest. Scientists work hard to chart the different types. Chart your own family using an anthill shape (like the one below). Write down the names of your brothers or sisters, your parents, your uncles and ants (oops—aunts!), and any other relatives you know!

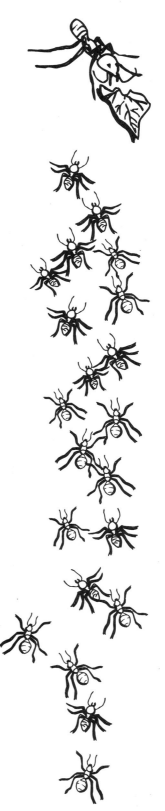

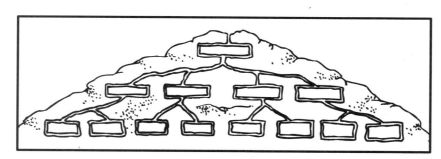

Just for Fun: Rain Forest Flip Books

Make a flip book! You'll need index cards, markers, and a stapler. Color the patterns, then cut them out and glue one picture to each card. (The squares go across, from left to right.) When you're finished, draw your own.

TOUCAN FLIP BOOK

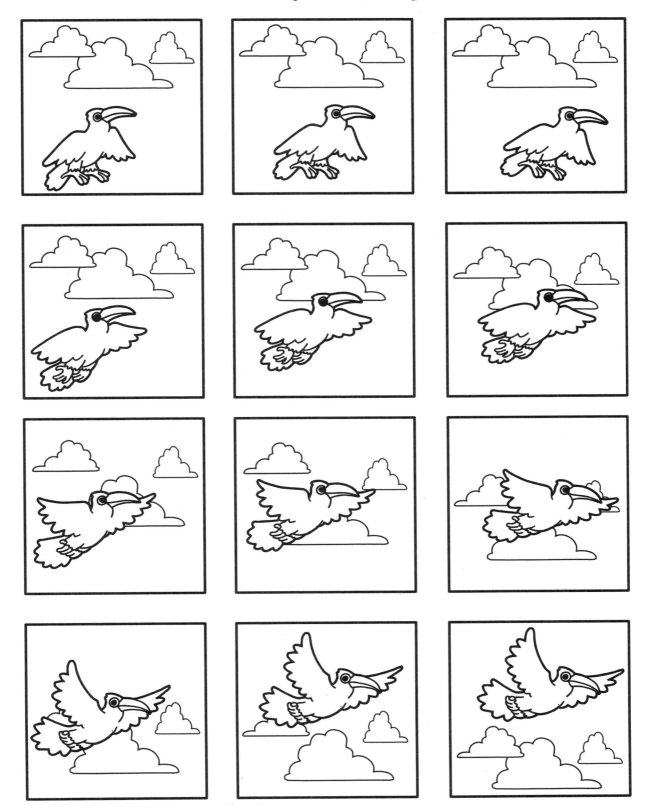

Monkey Flip Book

NAME THE ANIMALS

Some of the animals in the rain forest have very unusual names. For example, there is the okapi, the kinkajou, and the ocelot. Look at the animals below, then try to think up names for them. (The real names are based on what the animals look like. You can name them any way you want.) Check your answers with the real names (listed upside down) and then decide which name you like better, the real name or the one you chose.

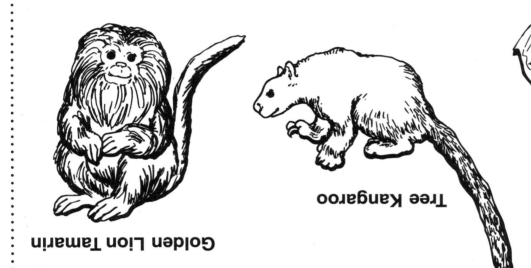

Golden Lion Tamarin

Tree Kangaroo

Flying Dragon

Hornbill

Just for Fun: A Rain Forest ABC Book

Use the Rain Forest ABCs to make your own ABC book. Choose one or more rain forest creatures (plants or animals) for each letter. Illustrate your choice. If you want, write one fact about the plant or animal you've chosen. Make one page per alphabet letter. Then bind your pages together in an ABC book. If you want, have a friend or two work with you on this project. Share the finished book with your friends or family.

RAIN FOREST ABCS

Army ants, ape, amaryllis, anteater, armadillo
Bat, bird-of-paradise, boa constrictor
Centipede, canopy, coral snake, chimpanzee
Dendrobates azureus (blue poison dart frog)
Eagle, epiphyte, emergents
Fungi, fruit bat, flying fox, fern, forest floor
Gorilla, gecko, gypsy moths
Hummingbird, howler monkey, harpy eagle
Insect
Jaguar, jabiru
Kinkajou
Leaf litter, liana, leaf-cutter ant
Monkey, moss, mudskipper
Nectar
Orchid, okapi, ocelot, orangutan
Parrot, praying mantis, porcupine, paca, piranha
Queen Alexandra's birdwing butterfly, quetzal
Red fan parrot
Sloth, spider monkey, silky anteater, slug
Tarantula, tree frog, toucan, tortoise, termite
Understory
Vines, vampire bat
X
Woolly monkey, wasp, water hyacinth
Yam
Z

Now I know my ABCs, next time won't you sing with me?

Note: Try to find rain forest words that fit the letters "X" and "Z." Or leave those pages blank in your ABC book.

SAVE THE RAIN FOREST

Today many rain forests are disappearing. About 50 million acres each year are vanishing worldwide. This is happening for several reasons. In some rain forests, the trees are being cut down for their wood. Other forests are cut down and burned to clear the land for farming, grazing, and building roads.

As the rain forests disappear, so do many species of plants and animals. For example, 40 percent of all birds of prey depend on tropical rain forests in one way or another. All apes, and most monkeys, consider the rain forest home.

If people continue to cut down the trees of the rain forest, many beautiful plants and animals will disappear.

Alert your friends and family to the need to save the rain forest. Create a newspaper listing information about different plants and animals that live in the rain forest.

Real newspapers answer the questions who, what, when, where, why, and how. Try to answer at least several of these questions in the stories in your tribune. For example, your "who" might be "a tree frog," your "what" might be "lived its whole life in a tree," "where" might be "in the rain forest," and your "why" might be "because it had sticky pads on its feet."

Just for Fun: Spread the Word

You can spread the word about saving the rain forest by making "save the rain forest" bumper stickers or stickers for book covers. Or send your friends Earth Day Cards. Earth Day occurs on April 22nd. It's a day when people are supposed to remember how important it is to take care of the Earth. Give your friends "Happy Earth Day" cards!

RAIN FOREST TRIBUNE

RAIN FOREST TRADING CARDS

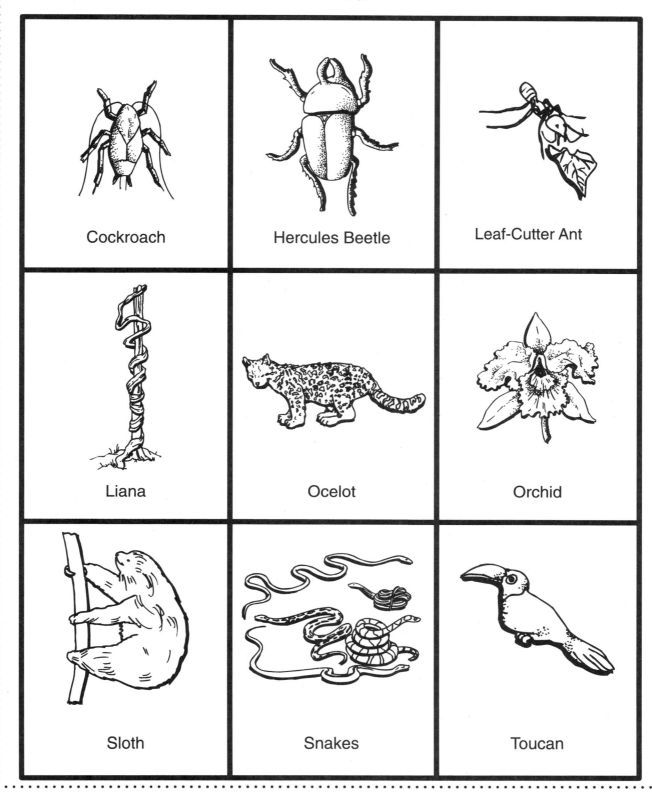

RAIN FOREST TRADING CARDS

Leaf-Cutter Ants: Leaf-cutter ants are hard workers. They snip pieces of leaves with their jaws and take the pieces back to their nests.

Hercules Beetle: This beetle is the size of a baseball! It has finger-sized pincers and looks like a knight in armor.

Cockroach: Most of the 4,000 types of cockroaches live in tropical rain forests. The Blaberus giganteus is as big as the palm of your hand!

Orchid: There are more than 25,000 kinds of orchids in the world! Three quarters of these live by perching on other plants.

Ocelot: The ocelot looks like a pretty kitty. However, this member of the cat family is a fierce hunter!

Liana: Lianas are woody vines. Some may grow to be even longer than a football field! In Tarzan stories, Tarzan swings from tree to tree on lianas.

Toucan: Toucans play games with each other. They even have food fights! They sometimes preen each other and offer each other food.

Snakes: Boas, constrictors, anacondas, and pythons are the largest snakes in the world. The first three live in Latin America, while pythons live in Africa and Asia.

Sloth: Sloths provide homes to different animals. Beetles, mites, and moths can live in a sloth's fur. One sloth was found with 978 beetles living on it!

Rain Forest Glossary

- **Buttresses:** The rain forest trees have shallow roots. This is because most nutrients are in the top of the soil. However, shallow roots make the trees less stable. Buttresses are thick supports that grow from the trunk. They help keep the tree stable.
- **Canopy:** The canopy trees grow so close that their branches can touch. The top of the canopy in most rain forests receives a lot of sun. The branches of the canopy block the sun from reaching the lower levels of the forest.
- **Ecosystem:** An ecosystem is a community in nature. It includes all of the living and nonliving parts. The rain forest is one type of ecosystem.
- **Emergents:** The emergent trees are the tops of the tallest trees in the rain forest. They grow taller than the canopy, poking through it.
- **Epiphyte:** Epiphytes are plants that live high in the forest. They don't have roots connecting them to the earth. Instead, they attach themselves with their roots to host plants and live there.
- **Forest Floor:** This is the lowest level of the rain forest. The fungi, herbs, moss, and other plants that don't need much sunlight live here. Only 1 percent of the sunlight reaches the forest floor.
- **Light Gap:** A light gap occurs in the rain forest where one or more canopy trees have fallen. This lets sunlight reach the floor. The gap lets in light. It can let smaller trees shoot up toward the sun!
- **Understory:** The understory is the layer below the canopy and above the forest floor.

RAIN FOREST RESOURCES

Books:
- *Conserving Rain Forests* by Martin Banks (Steck-Vaughn, 1989).
- *Explore the World of Exotic Rainforests* by Anita Ganeri (Ilex, 1992).
- *Journey Through a Tropical Jungle* by Adrian Forsyth (Simon & Schuster, 1988).
- *Jungle* by Theresa Greenaway (Dorling Kindersley, 1994).
- *The Living World: Jungles* by Dr. Clive Catchpole (Dial, 1983).
- *Nature's Green Umbrella: Tropical Rain Forests* by Gail Gibbons (Morrow, 1994).
- *One Day in the Tropical Rain Forest* by Jean Craighead George (Thomas Y. Crowell, 1990).
- *Rain Forest* by Barbara Taylor (Dorling Kindersley, 1991).
- *Rain Forest Babies* by Kathy Darling (Walker, 1996).
- *The Rain Forest* by Billy Goodman (Tern, 1991).
- *Shrinking Forests* by Jenny Tesar (Facts on File, 1991).

Web Site:
- Animals of Peru's "Manu" rain forest
http://www.pbs.org/edens/manu/flora.htm